Contents:

PRESENTATION: THE FUTURE OF SPACE - the Road is Paved.

John David Hanna

In accordance with the mandate of the 2nd International Space conference - "this conference and those to come must serve the best of our youth by engendering a plan for the birth of a pan-human society spanning the solar system and beyond" - **and true to the data researched** this presentation will demonstrate that alien machines have smoothed our path to the stars.

Introduction

While you read this booklet it will become obvious that the path already exists but the question remains as to who it was built to service. Is it for us to use and others are watching our progress? Is it for us and the others are so advanced that they know what we will do? Was it build entirely by machines for another? If for others was it done awaiting the arrival of the aliens that never arrived? Did these aliens arrive and die out or move on? Are we the aliens or are theses aliens our ancestors?

None of these questions really matter – **we are here and this is the path we are on.**

It isn't necessary that you believe that someone made the path. Maybe it was a coincidence. This presentation concerns 'the path' and how we will follow it onto the stars whether it is rife with ancient relics or not.

The path begins on earth with just enough oil to get us off the planet yet will run out before we poison ourselves. As we begin to worry about atmospheres, a necessity for space travel, we move manufacturing to the moon and orbiting habitats. There, simple solar furnaces provide nearly free and unlimited energy. We will devise a way to return it cleanly to earth. Lacking carbon on the moon we can find that element among the asteroids along with enough rare materials to satisfy every need. The moon economy will become so robust it will change our earth economy from scarcity profit to abundance distribution. Isn't that interesting because you might imagine that to be the type of economy needed on a starship or a habitat.

We will adapt to space in many ways. Biologically to cope with more or less or no gravity. Technically for food production and maybe gravity controls. Psychologically to avoid animal behaviors that could destroy a habitat of trusting neighbors. Radiation will always need to be shielded.

Later, as we power our habitats for star travel we will need to manage those ultra-powerful thrust beams and those out-flung gas giants will make excellent heat sinks. Saturn is something else! Doped ice rings for exterior structure construction and hydrogen for fueling in-system. A walled off north pole provides easy and efficient access to the valuable gas. Variant moons surround the planet for every minority need – hydrocarbons for the food printers and plastics, carbons for internal structure. Mini-moons abound in ancient orbits stored like planes on the tarmac.

Aliens or Alien Machines?

The Drake Equation [1] is acceptable to most scientists, as a probability proof, that new alien civilizations come about in this galaxy at the rate of 10 per century and that these civilizations last for millions of years. Other scientists say that number is overly large and that civilizations last for far less time (although they may evolve into something greater instead of destructing). The Drake Equation says that other civilizations probably have existed, probably do exist, and probably will come into being but it does not say that we will meet one.

If this world is to interface with other sentient beings, being that the Drake odds are very low, we will further need to beat the odds of time. Our civilization will need to be developing along the same timeline as theirs. [2] When we meet them what if they are just too ugly, smell horrid, or have a very odd trait like coprophagia? [3] Aliens may exist but the odds are increasingly bad that we will get along with them.

The odds can be increased if we multiply by the number of machines they send out. What if, in the future, we send out self repairing, self replicating intelligent robots that can transform an entire solar system to make it ready for our arrival. There is no cost; the robots use local energy and refine local ores and manufacture whatever is needed. They would do one then another even if we don't arrive in person. If they/we send out just one of them and they last a long time that would significantly **increase the odds of interfacing from a rarity to likely.**

The Moon as the Next Step

Iron, aluminum, copper, zinc, sulfur and titanium abound on the moon. [*5] In most industry energy is a major expense but on the moon it is negligible, almost free. "At an intensity of 1366 watts/m2, the lunar surface is ideal for the use of solar furnace technology." [*6] Temperatures high enough to drip metals from the regolith ore and free oxygen, up to 20 percent of the mass, can easily be generated. Solar furnaces could also substitute for the radioactive boilers of a nuclear power plant providing inexpensive electricity if the heat exchanger, generator and associated controls are transported.

NASA has a handle on the abundance of the moon. Using the solar furnace to melt the regolith connected to a generator enables the use of electrolysis on synthetic regolith.

With this method the scientists have successfully tested metal and oxygen extraction [*7] with astoundingly good results. Thin film solar cell production is also viable as a moon industry taking advantage of the proliferation of the silicon salts and the native vacuum. The cost of power generated on the moon in the proffered papers is projected to be higher than the cost of earth generated power however, moon power generation cost has proven, in newer theory, to be negligible. [*8] Solar cell delivery from the moon to the earth is feasible with a simple freight orbit and an electrical mass driver to propel the materials off of the surface. [*9]

The Moon as the Next Step (cont)

If the product could be encased in a lunar manufactured glider that can withstand earth reentry the retail cost on earth could be further lessened.

Solar cells are a superior product if removed from the pollution and energy required to produce them – superior to any of the newer alternatives now in early usage. Microwave beams from space may work well enough but many people fear an accident or an intentional confiscation for militant purposes. Actually solar cell production could benefit from moon vacuum and it seems reasonable we should endeavor to maintain that vacuum. [10]

At first over-thought a moon base does not appear to be the miracle of invention it once was deemed to be. The ESA has an excellent concept which may well deploy in the near future. [11] For moon colonies a few amenities would be needed to make life livable for the colonists. Aside from the cavernous lava(?) tunnels a surface structure can take advantage of the nature of the regolith and the abundance of oxygen and metals. Extract the oxygen for the atmosphere, the metals for structural support and compress the slag into bricks.

Robots could construct walls ten feet wide, 30 foot high and miles long – some areas of 'Transition' glass would provide a view. Fill the area with sustainable ecosystems obtaining carbon from an asteroid and voila – home away from home.

Phobos

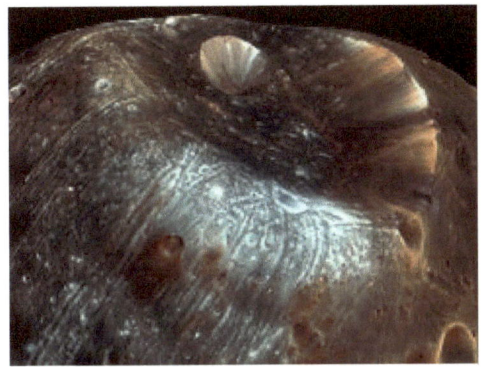

NASA has shared some excellent pictures of the Martian moon but three Russian probes have failed. Suspiciously the NASA imaging probe never intended to closely approach the moon while the three failed Russian probes plotted a collision course. The colorized NASA MRO photo of Stickney Crater 2008 is shown aside. [*12]The characteristic measures have changed over the last decade and tend to discredit the unnatural label provided by the Russian Space Agency while the ESA is back and forth on their declaration. For instance the density has moved from 1.08 to 1.88 g/cm3 which remains near a third the density of dirt.

Lowell explained his observations of the martian canals as evidence of civilization. Although very reasonable this is a clear warning concerning extrapolation of incomplete evidence. This paper will take the position that Phobos is part of the 'path' and expound on the abnormalities if and until proven otherwise.

In 1988 the Russians reported witnessing an outgassing from Phobos although no evidence of volatile molecules is seen on the surface. The ESA has concluded that Phobos contains large voids inconsistent with a captured asteroid. Several radar mappings of Phobos over decades have been analyzed and reanalyzed with the same result. Is it live? Was the outgassing an orbital correction?

(Phobos)

- Phobos orbit eccentricity is 0.0151 (perfect circle) and a capture would very likely be more eccentric.

- Not only is the orbit a perfect circle, backward, but it orbits directly over the equator. A prograde moon might have formed from the planet in such a manner but a capture could be from any angle. Such a perfect capture is impossible.

- Orbit is retrograde and a capture would tend toward prograde elliptic. It's supposed origin, the solar ecliptic, is 27 degrees axial tilt from the planet Mars so why isn't it still circling as it was when supposedly captured.

- Phobos orbits the planetary equator quickly and closely putting every particle of the planet within range of possible sensors every few hours. Phobos completes an orbit far quicker than any other solar moon system.

- Because of the low density some postulate a loose gravel composition while others declare it is more like a hollow canister and exhibits eccentricities closer to a can. If Phobos were natural it would not be able to survive the impact that created the Stickney crater.

- Stickney crater can always be seen from the Mars surface but not directly facing.

- Radiating grooves and crater chains can be seen extending laterally across the surface. The marks look like grapple and ordinance scars.

- Although postulated to be very old there is no determination of when it was 'captured'. It is most likely dead or automated or partially automated – there are no living beings on or within it unless they are hibernated.

- Unfortunately this moon is unnatural **and it appears to be a weapon.** It is very old and we can hope it is not functioning at full capacity. We can hope it is assigned to Mars and is not a gatekeeper or concerned about anything other than Mars.

Asteroid Belt

Further out on this paved road we have the asteroid belt with it's repository of riches. Once mining techniques are developed and in production the platinum group metals from one minor metallic asteroid, one of millions, would produce three times the gold ever mined during the entire history of the Earth. [13]

There isn't really anything to report about chunks of various materials laid out for the use of habitats, ships and colonies.

If it were done purposely it couldn't be more convenient and profitable. Supply robotic drones and factories of various forms to redo this raw material into needed products and luxury items will soon abound. This embarrassment of riches will further draw society into the shared socialistic governance of inevitability.

As a far fetched aside, these rocks could be cold storage for aged individuals or groups of individuals in stasis. It was mentioned and will be further illuminated that once the extended lifespans of 'the others' is finished they may elect to slow or stop their lives temporarily. If this were the case their actions would be so slow as to be unnoticeable, perhaps as slow as one second per our day. It's certainly cold enough out there to save on the power.

Saturn

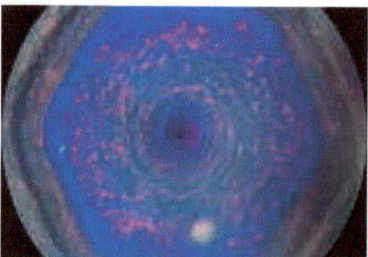

What a cornucopia of materials is the Saturn system with it's moons and rings – it's like a candy store for children. The assumed rocky inner core of this gas giant swept up quite a bit of gas to form this low density giant that ends up with one gee of gravity. Even if swept up wasn't the way it happened the fact is it has a lot of hydrogen gas.

If you need fuel for your inner planet voyage this gargantuan has an ample supply. If the northern retaining walls of unknown origin remain stable we could land a ship there – it's only one gee! We might be able to ride the one way 500 mph wind to lift our gas loaded ship out – very economical! We could find out what forces created those hexagonal walls [14] and that would be a boon to space geologists or for relic seekers.

Saturn Rings

Although the rings are mostly water ice they are as diverse as the moons. The median particle size is segregated sharply by the ring boundaries and they have various trace amounts of tholin, carbon, and oxygen as revealed with spectro-scopic analysis. [15]

Steel is made with iron and precisely measured aggregates or carbon, nickle and other elements. It is postulated that water ice changes state at very low temperatures to a metallic lattice similar to stainless steel at STP. [16] A star ship would operate in a cold environment and steel ice may be useful for exterior structure. Even as ablative protection plating or mass to feed to the ionic engines it couldn't be better than the materials of the rings. Atoms between stars are so very rare that a ship should survive even at percentages of light speed.

(Saturn rings)

In a star ship with a large ion drive structural or ablative plating would likely be cannibalized for the plasma drives and would need replenished at port. Here it is, all cold and laid out in various flavors all easy to obtain. Why would anyone need to pay for materials provided by self repairing machines whose costs, over time, limit to the infinite zero?

There is agreement, because it can be clearly seen, that the **shepherd moons** are <u>responsible for the maintenance</u> of the rings. The how and why of their operation remains unexplained. Prometheus and Pandora orbit together, one inside and one outside the F ring of Saturn. As they move a force similar to a boat's wake can be seen affecting the particles. It seems natural, as of nature, until reason surmounts and facts are reviewed. The particles effected are many tons of mass while the shepherds are less than one hundred miles in diameter – so it isn't their respective gravities that cause the disturbance. There isn't any medium (air, water) to carry a wake.

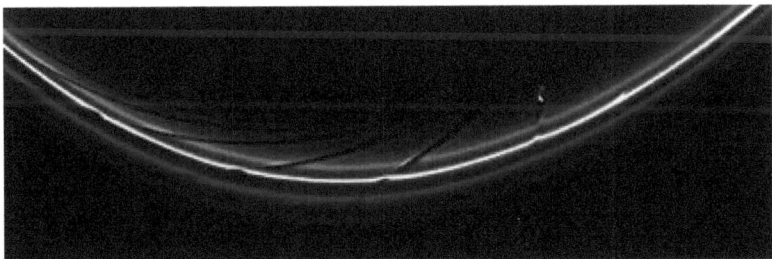

The obvious solution is some type of gravity tool. Perhaps a similar machine was used to help we natives construct those massive stone works we see around the planet. A gravity wave projector, wouldn't that be a lovely thing to have. Also a mystery is how materials of various sizes are segregated to differing rings like piles of gravel of various widths at a cement yard.

Gas Giants

There are storms on all the gas giants yet they have no life, probably. **SO** if you are landing a moon rocket in Florida you might have to interface with Florida civilization – there are rules.

If you were driving a star ship inbound or outbound you are probably using a pretty large motor and you wouldn't want to burn the atmosphere off one of the inner planets by accident.

When we do light one of our own starships we may decide to use these planets as heat sinks. How much scarring would a planet suffer under the blast of one of these giant engines? The flame would probably penetrate clear through the gases and produce irregularities on the heavier elements below. Would mountainous scars torn into the subsurface result in visible storms like the eye of Jupiter?

Starship exhaust would likely be invisible. The designer would want every percent of power for propulsion so there would be no leakage. It would be a strait laserish beam of atomic pieces boosted to near light speed or beyond. Although miles wide and fiendishly hot you would only be able to detect it if you were unfortunate enough to be under it. The beams may well be lethal for light years of their length.

Klerksdorp and Iapetus

Klerksdorp is a town in South Africa, rather a mine, where odd metallic globes are embedded in soft rock that resemble the Saturnian moon Iapetus, shown below side by side.

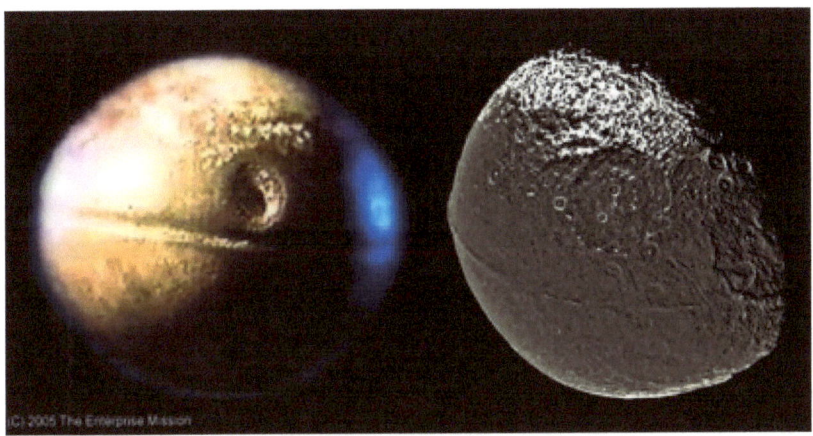

Sitting in the museum of Klerksdorp, South Africa, hundreds of tiny metallic balls (known as the Transvaal Spheres) are displayed. These spheres have been found by South African miners during the last several decades. [*18]

They are a blue metallic encased in sedimentary rock. Sometimes called cosmic cannonballs due to their resemblance to Iapetus. Perhaps they were layered with life creating virus and set out upon the solar system from the newly built moon?

But the alarming aspect to the story is that these objects are found in rock dating back to the Precambrian era – around 2.8 billion years ago.

Roelf Marx, curator of the Klerksdorp museum said:

"the spheres are a complete mystery. They look man-made, yet at the time in Earth's history when they came to rest in this rock, no intelligent life existed. They're nothing like I have ever seen before" (Jimison 1982).

Phoebe and Iapetus

Iapetus and **Phoebe** orbit in opposite directions, Phoebe retrograde, yet they have many similarities. These two moons are composed of metal and carbon according to the spectral analysis [15] while the other moons, 62 at last count, are various water, methane, tholin and rock. [17]

When planets were formed, according to the current theory, star dust coalesced from the effects of gravity. The dust began to spin as the particles fell into the sun. Outward, islands of mass formed and collected their own dust to become the planets all swirling in the same direction as the initial particles. Eventually everything is settled outward from the sun's equator, like a plate.

You would think that planets would be in the elliptical line themselves but instead they vary as to their tilt. It is what gives the earth seasons. The earth is tilted at 23 degrees from upright association with the solar ecliptic. The earth's moon is in line with this solar ecliptic and does not coincide with the Earth's equatorial plane, as most moons of other planets do.

Both of these Saturn moons are inclined out of the solar ecliptic and equatorial ecliptic, Phoebe to the south. Both moons have very low density, less than that of Phobos. If it should become time to set up a control tower for a Saturn starport between these two moons the entire area is in view.

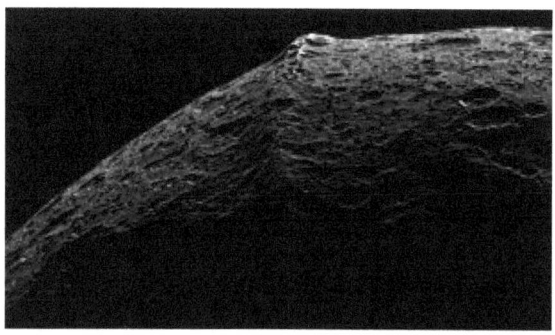

The equatorial ridge of Iapetus is pictured at a tilt. The ridge extends around seventy percent of the moon located exactly at the equator. It is up to 13 kilometers

Phoebe and Iapetus (cont)

high and 20 kms wide. It is covered by impact craters indicating it is as old as the moon and not a recent addition. This type of structure has only been seen on this one moon. Many explanations have been proposed for this structure but all are stretching the limits of believability and probability. The obvious theory is that it is a seam joining the two halves.

Other pictures reveal a geodesic pattern with plate edges that span over one hundred miles. Government scientists are frantic to announce (in my opinion) that the carbon iron surface is a dust covering loose ice and certainly not steel plates with a hollow interior. Like Phobos, the density could be indicative of a hollow, spherical, iron plated habitat or loosely packed gravel with craters so large as to have destroyed a light material.

Iapetus has a very low radar albedo meaning it reflects little back from ground penetrating radar as would be the case of a hollow structure or interior 'impurities' that scatter the beam. Of all of Saturn's moons, the only radar signature lower than Iapetus was Phoebe. [*19]

The large crater shown previously, with the Klerksdorp spheres, Turgis [*20], is 520 km in diameter and 40 percent of the diameter of Iapetus. It is located front and center. An even larger impact crater Abisme is located to the right and looks like the energy projector of the 'Death Star' of 'Star Wars' fame. It's a waste of time to speculate on the usage, if any, for these two depressions. One thing is certain: although most scientist repeat the crater mantra there is no way that a gravelly icy ball could have survived an impact from another body of those sizes. There are no bumps when two bodies meet in space. Impact speeds vary between 20 miles per second up to 120 miles per second! [#21]

*The following sections
Soundings, Timing and Hard Code
contain my opinions
and less documentation.*

Soundings

The SETI institute has been searching for alien signals for decades to no avail that we are aware of. The 'WOW' signal was detected by Jerry Ehman on August 15, 1977. The 72 second signal at 1420.4556 MHz was received on what is known as the hydrogen band which is normally quiet of background radiations. It was never repeated and is acknowledged that it could have been leaked from a source much nearer – a prank.

However, SETI is not now monitoring the hydrogen line itself but harmonics of it. For instance 2 x 3.14(pie) x 1420.406 = 8.92 GHz. This two pie line signal can not be a natural radiation.

Radio emissions from Neptune and the other gas giants were first discovered by Voyager. These very low energy signals were near the 600 KHz spectrum. They recurred every 16 hours and were used to determine the unknown rotational period of the planet. These dozen or so emissions were only detectable when Voyager was nearest to the planets surface. [*22] No decoding was attempted and it isn't known if the constant signal repeats the amplitudes, like an SOS signal repeating.

There is less than a dozen of these emitters on this planet and they are very low power. Why projections from pinpoint sources? If it were a natural occurrence it would seem that something the size of a planet would have many, many sources or none. What else could it be; shipwreck beacons? What about information like those AM broadcasts along the highway when we approach sites of interest?

Pinpoint signals, only detectable on solar approach, have been discovered at many places in the solar system. Comet 67P provided a signal for the Rosetta spacecraft as it approached. This signal repeated itself on a magnetic oscillation that seemed totally unexpected.

Timing

We are making progress with suspended animation. It was discovered that a person could live for hours without a heartbeat or a breath. These persons have to be cold but medicine is already using the process to deanimate persons so there is time to perform needed procedures. [*23]

Metabolic flexibility is being investigated so that people could be deanimated using hydrogen sulphide. This chemical is found in the brain and determines the amount of oxygen the body needs. The body can be fooled to think it needs less. Once deanimated (dead) they can be placed in cold suspended animation for a long time and returned without harm. Dogs have been dead for weeks and returned unharmed. This is DARPA funded and their progress is classified. It is thought to be a viable solution to deep space exploration.

Freezing damages the cells so it can be speculated that the imagined subject would be kept just above freezing. The body and mind would be very slowed but theoretically a person could remain in this state for centuries before being thawed in the same condition as they were when they began their sleep.

Perhaps wiring can in melded with the cold body that can access the very slowed brain. Perhaps inputs can be accessed and the brain can receive real world news, consolidated and re-timed by computers. Perhaps mechanical extensions can be connected controlled by the brain with computer interfacing. The machines would be enabled in both the fast and slowed worlds.

Imagine that one slowed body could communicate easily with another slowed body. There wouldn't be any difference of timing between the two, or between a massive community of the slowed. If the slowed bodies experienced one second per our day they could experience traveling to another star, using existing technology, in a few hours (by their experience). They could brunch with their buds at different stars every few days.

(Timing)

Our cold selves could have a choice between virtual world connections or the real world connection. In the real world a computer will be needed to adjust times. Not so in the virtual worlds where we can create our own avatar to interface with others.

But what about aging of our actual bodies? New research has been done with telomeres with promising results. [24] They can be rewound while limiting the cancers. In the future before we elect to slow ourselves down instead of death, we will want to get all the usage we can from our normal bodies.

How long do you think we can live when the aging process is reversed and rejuvenation is standard. Will it start at a lifetime of 300 years? How long until we live to 3000 or 30000 years? My guess is with long life will come risk aversion – when you can live to be 26576 years of age who wants to die at 67?

Cold storage will have a down side assuming the body can be thawed returning to normal time. It will take awhile to defrost and might even hurt. A good place for storage would be where it is already cold, very cold. The outer solar system or even the Oort clouds would be ideal.

Only appealing to avoid death, perhaps, what about transfer of intelligence into a computer? The artificial body of real time, attached to the real body in the cold sleep, could be animated among 'us' and as handsome or beautiful as wanted.

The actual brain has much to influence it. There are about 1000 connections to each neuron and each one can carry 1000 different analog signals. Besides the decisive complexity each part of the cell and even the synapses can be changed by myriads of conditions in the surrounding chemistry. **There isn't enough time** to access the data layers before the brain dies. On a slowed brain one slice at a time, over perhaps thousands of years, might be possible. The possibilities for an electronic being are endless.

Hard Code

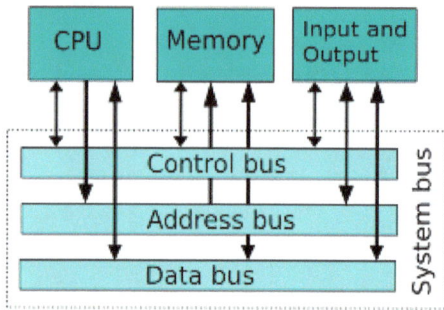

Doesn't it seem like the corpsicle route is like a logical progression? It explains where all the aliens are and why this solar system is so littered with relics. What really bothers us is that we are here on this lovely glowing marble of a planet and nobody is bothering us!

How could life be more adapted to space than if we needed the cold of the outer solar system because it was easier to maintain temperature? Our biological lifespans, even enhanced to 30000 years, are a flicker of the eyelid compared to the ages we can encounter when the sampling rate is adjusted to one second per Earth day!

So it seems that every once in a while, thousands of years between instances, things happen in the DNA. Only 8000 years ago dogs didn't exist, only wolves. You might get weenie wolves to branch out to chihuahua by selective breeding but I just don't see it. As a species we were pretty weak back then and the helpful dog could have made the difference in the survival of us.

OK, the aliens did it; but what kind of alien hung around and was so interested in 'us' that it sacrificed that many years in utter boredom to watch over us? If there are aliens about there must be better places to be than here with us.

These slowed people did it, that's who! They could be in cold cities on Titan, Phoebe, Iapetus, on Saturn, in Neptune or encased in asteroids. They don't move, they communicate with undiscovered methods (and very slowly) and interact with artificial intelligences at their command. The AI can take many shapes and protrude any number of interfaces be it chemical, biological, physical or psionic.

(Hard Code)

So these artificial intelligences (AIs) are about doing duties for the cold masters. Most of the time the machines are on their own because of the time differentiation and the condensed manner in which they update the dead. They probably don't have a compassion module and simply calculate facts and effect change accordingly. For instance, war moves a civilization ahead very swiftly but if the people were given a choice such conflicts and loss wouldn't happen.

Perhaps we are moved on whim or perhaps we have an amiable purpose or an unknowing slaved purpose. Someone or something with experience designing civilizations for their own use would design them in an economical fashion. The design would be done in such a way that the designed would want to do the mission that is planned out for them. Do you want to be an astronaut?

Perhaps some of our own people are collaborators with an alien interface but if so it is unlikely that they know what the outcome of their actions will be. It is unlikely they know whom they are working for. The idea that they were working with aliens would provide a strong incentive for them to do as instructed.

There is hope and evidence of compassion. For one thing, we exist without obvious whips and chains. We aren't damaged for glee or on a daily basis. Most importantly of all, our hosts are civilized.

Civilized means there are laws. Laws to deal with the clueless natives. People say 'If aliens constructed some of the stone constructions here on earth why were they confined to just stone?' The answer could be that it's the law. When dealing with a growing race you don't provide implements beyond their industrial level as popularized by Star Trek. Why? Because it discourages them (us), makes them feel small and insignificant.

Do civilized alien races know how best to bring us into the 'confederation of intelligences'? If it exists, probably. But this means don't expect a saucer to land on the front lawn of a government building. Do they treat us inhumanely? Not really but our care may be turned over to robots until we are wise enough to contribute.

Final

It doesn't matter whether you believe that the path was laid by others for us or laid for others and we inherited it. What matters is that the path exists and that is the way we will progress to interstellar space.

This is my booklet and I am allowed an opinion as long as it is clearly marked as such and the opinion below is my opinion and subject to change.

We have recently been awakened by an AI messing with our DNA. I don't know what the mission is. The AI is a machine and does not have a human interface, it is logical. War is logical, it moves us forward quickly for the greater good but the price paid by some individuals is ruthless. Hopefully the perceived need will soon end.

It may not matter that we are made to serve a machine. We still have love, beauty, companionship.

Further, when designing a living machine, the designer would receive better service if the construct liked what it was tasked to do.

(Finally)

So where are these aliens?

- Aliens exist in normal and expanded time. The real time peoples are most likely migrated toward the center of the galaxy where the history and action is. The cold ones may be all over the place. They will have their protections.

- The normal populace didn't expect comet 67P to be occupied but it sent out a signal as we approached; normal procedure most likely. It was like a plane nearing an airport. Then it disabled any machinery that would harm the comet's hull while allowing pictures and observations. It most likely read the hard coding and then disabled that which would disturb their craft.

- Why is the craft so ugly? Probably stealth. There may be stupid berserker crafts roaming the stars that are fooled by limited methods – as we are but we're still new.

- Phobos has a similar history of defense. Three Russian probes suffered software failures and all three were going to land. The last one destructed as soon as it left the atmosphere and set it's course (for Phobos). If planetoid or habitat defenses were run by advanced robotics of course they would have methods of dealing with robotic threats.

- Kepler had to resort to a secondary plan when the gyroscopes developed problems. The degraded mission continues but the delay could have provided enough time for the code to have been completely rewritten onboard, unknown to us, and blindspots could have been inserted.

Yes, blindspots are a stretch. If anything be heartened that, although we appear to be immersed in a traveled area, the occupants know what they are doing and don't seem to be planning some type of retaliation or some type of torture.

References and Data

1 http://www.space.com/22648-drake-equation-alien-life-seager.html The Drake Equation Revisited: Interview with Planet Hunter Sara Seager By Devin Powell, Astrobiology Magazine | September 04, 2013 06:

2   8 000 000 000 000 years of Earth / 10 000 length of civilization => 80 000 000 one chance in 80 million that a nearby civilization will develop at the same time as ours

3   https://www.princeton.edu/ ~achaney/tmve/wiki100k/docs/Coprophagia.html The article content of this page came from Wikipedia and is governed by CC-BY-SA.

4   http://www.scientificamerican.com/ article/rise-of-the-robots/ Rise of the Robots – The Future of Artificial Intelligence Mar 23, 2009 By Hans Moravec .

5   http://www.space.com/13247-moon-map-lunar-titanium.html Moon Packed with Precious Titanium, NASA Probe Finds SPACE.com Staff | October 11, 2011 07:00am ET

6   http://www.lunarpedia.org/index.php?title=Solar_Furnace Solar Furnace JA Rogers, August 2008

7   http://isru.nasa.gov/Molten_Regolith_Electrolysis.html July 3, 2012. Page Editor: Laurent Sibilleb NASA Official: William Larson

8   http://ntrs.nasa.gov/archive/nasa/casi.ntrs.nasa.gov / 20050110155.pdf Manufacture of Solar Cells on the Moon NASA Cross Enterprise Technology Development Program

9   http://en.wikipedia.org/wiki/Colonization_of_the_Moon Colonization of the Moon various authors, 2013

10   http://www.geoffreylandis.com/moonair.html Air Pollution on the Moon by Geoffrey A. Landis 1990

11   http://www.esa.int/Our_Activities/Technology/ Building_a_lunar_base_with_3D_printing Building a Lunar Base with 3D Printing 2013

12   http://en.wikipedia.org/wiki/Phobos_(moon) Phobos

13   http://www.reuters.com/article/2012/04/24/ us-space-asteroid-mining-id USBRE83N06U20120424Tech billionaires bankroll gold rush to mine asteroids by Irene Klotz Apr 24, 2015

14   http://ciclops.org/view/5950/Spring_Unveils_Saturns_Hexagon?js=1 Spring Unveils Saturn's Hexagon

15   http://arxiv.org/pdf/1203.6230v1.pdf Saturn's icy satellites and rings investigated by Cassini – VIMS. III. Radial compositional variability

16   http://militzer.berkeley.edu/papers/ice26.pdf New Phases of Water Ice Predicted at Megabar Pressures Burkhard Militzer1, 2 and Hugh F. Wilson1 1Department of Earth and Planetary Science 2Department of Astronomy, University of California, Berkeley

17   The Moon of Saturn http://en.wikipedia.org/wiki/Moons_of_Saturn

18   https://www.forbiddenhistory.info/?q=node/26 "Forbidden Archaeology", M. Cremo and R. Thompson (1998)

19   Cassini RADAR observations of Enceladus, Tethys, Dione, Rhea, Iapetus, Hyperion, and Phoebe. By M.A> Janssen etal., April 2006 https://www.researchgate.net/publication/222218843_Cassini_RADAR_observations_of_Enceladus_Tethys_Dione_Rhea_Iapetus_Hyperion_and_Phoebe

20   Turgis (crater), https://en.wikipedia.org/wiki/Turgis_(crater) January 2016

21   http://astronomy.stackexchange.com/questions/8707/ how-gently-could-a-comet-asteroid-meteorite-hit-earth, Astronomy beta 2015

22 Neptune and Triton, D.P. Cruikshank editor, University of Arizona Press, 1992.

23 Ted Talks, Mark Roth Suspended Animation, 2010

24 Stanford Medical News Center, Telomere Extension Turns Back Aging Clock in Cultured Human Cells, Helen Blau, January 2015

Party Jack

John David Hanna authored the Hard Jack series of science fiction. A computer engineer familiar with expert and artificial intelligence systems, their strengths and abilities, those fictional books detail a roadmap to the requirements necessary to meet and greet the AI that runs this solar system.

John is an advocate of space exploration and believes the time is now for private venture.

Beg and borrow for moon construction money.

John has written the following science fiction titles.
Sexual incidents occur:

Hard Jack
Hard Jack Junior
Hard Hack III
Party Jack (short story)

Recently released, PG 14

Moon Jack

It's time the people that paid the bill for space exploration gain access to their own information. John attends international space conferences as a public service. John is a member of the IEEE and will present this researched information there in the near future.

There is no reason for the bias launched against any mention of 'others'. The disdain is a result of constant propaganda assailing the American people for seventy years. It is time to let the people handle their own destiny.

Think Big !

www.ingramcontent.com/pod-product-compliance
Lightning Source LLC
Chambersburg PA
CBHW040820200526
45159CB00024B/3082